2024 TESLA MODEL Y CAR REVIEW

Your In-Depth Guide to the Car's Interior Features, Pricing, Model Y Variants, EV Technology, Driving Experience, Safety Features, Performance, Pros and Cons

JAMES AUTOTRENDS

Table of Contents

Introduction

Overview of the 2024 Tesla Model Y

The 2024 Tesla Model Y, an evolution of its predecessor, the Model 3, stands as a formidable player in the electric SUV market. Retaining Tesla's distinctive sleek design, the Model Y brings practicality to the electric vehicle (EV) landscape with its high-riding body style. It shares its platform with the Model 3, providing a foundation for a roomier cabin and increased cargo space. While the transition to an SUV alters its handling dynamics compared to the Model 3, the Model Y compensates with enhanced passenger comfort and a more accommodating cargo area. The SUV boasts an impressive estimated driving range, with the Long Range model reaching up to 330 miles on a single charge, outclassing many of its rivals in the EV segment. As the automotive industry shifts towards sustainable alternatives, the 2024 Tesla Model Y emerges as a compelling choice for those seeking an electric SUV that balances range, performance, and practicality.

Notable Changes for 2024

As the 2024 model year unfolds, Tesla introduces subtle enhancements to the Model Y, reinforcing its commitment to innovation and refinement. While specific changes for this model year may not be officially announced due to Tesla's unique approach to updates, industry insiders anticipate a styling refresh codenamed Project Juniper. This refresh is expected to bring aesthetic improvements, though details remain speculative until an official release. Tesla's mid-year alterations occasionally surprise consumers, adding an element of unpredictability to the brand's evolution. Despite this, the 2024 Model Y maintains the core characteristics that have contributed to its popularity—impressive driving range, electric performance, and the allure of Tesla's cutting-edge technology. As the electric SUV market continues to evolve, these potential updates position the 2024 Tesla Model Y as a forward-looking contender, poised to adapt to the ever-changing landscape of electric mobility.

Pricing and Trim Options

Navigating the world of electric vehicles often involves weighing the balance between cutting-edge technology and affordability. The 2024 Tesla Model Y, expected to be priced between $46, 000 and $55, 000 depending on the chosen trim and options, continues Tesla's tradition of offering compelling EVs at various price points. With three trim levels—Standard Range, Long Range, and Performance—buyers can tailor their Model Y to suit their preferences. The Standard Range, starting at an estimated $46, 000, provides an entry point for those seeking a cost-effective electric SUV. The Long Range and Performance models, priced at around $51, 000 and $55, 000 respectively, offer enhanced driving range and performance. Tesla's transparent pricing strategy, coupled with the potential for government incentives, positions the 2024 Model Y as an enticing option for those ready to embrace the electric future without compromising on features or performance.

Chapter 1: Driving Experience

Performance and Acceleration

The 2024 Tesla Model Y excels in the realm of electric vehicle performance, showcasing the brand's commitment to delivering exhilarating driving experiences. The Performance model, in particular, stands out with its dual motors that generate a combined output of 470 horsepower, catapulting the SUV from 0 to 60 mph in a swift 3. 6 seconds. The Long Range variant, while slightly less powerful, is no slouch either, achieving 60 mph in just 4. 4 seconds. The instant torque delivery characteristic of electric vehicles contributes to a seamless and rapid acceleration, providing a sensation of continuous power throughout the speed range. Whether navigating city streets or merging onto highways, the Model Y's performance capabilities offer a dynamic and responsive driving experience, reinforcing the appeal of electric powertrains beyond environmental considerations.

Handling and Fun Factor

While the Model Y inherits its platform from the sportier Model 3, the transition to an SUV body style alters its handling dynamics. The result is a driving experience that, while competent, doesn't match the sheer fun and agility of the Model 3. The higher center of gravity introduced by the SUV body leads to a less spry feel in corners, and the ride can be considerably rougher over road imperfections compared to its sedan counterpart. Despite these compromises, the Model Y still offers a confident and secure handling experience, with precise steering and a smooth ride on well-maintained roads. The compromise in handling is offset by the increased practicality of the SUV, making it a suitable choice for those prioritizing passenger and cargo space over outright sportiness.

Driving Range and Efficiency

One of the standout features of the 2024 Tesla Model Y is its impressive driving range, positioning it as a leader in the electric SUV segment. The Long Range model boasts an estimated 330 miles on a single charge, providing a substantial buffer for everyday commuting and long-distance journeys. Even the Performance model, with its slightly reduced range of 303 miles, remains competitive in the electric vehicle market. Tesla's focus on maximizing efficiency contributes to these impressive figures, showcasing advancements in battery technology and aerodynamics. For consumers with range anxiety concerns, the Model Y offers a reassuring solution, demonstrating that electric vehicles can deliver not only on performance but also on the practicality needed for daily use.

Real-world Driving and Fuel Economy

As electric vehicles continue to integrate into daily life, understanding their real-world performance and efficiency becomes crucial. The 2024 Tesla Model Y, in real-world driving scenarios, provides a seamless and quiet experience, characteristic of electric powertrains. The cabin remains well-insulated, ensuring a tranquil environment for occupants. However, it's important to note that factors such as driving conditions, temperature, and individual driving habits can impact the actual driving range. In terms of fuel economy, the Model Y, measured in MPGe, reflects the efficiency of electric propulsion. While the EPA estimates are impressive, real-world results may vary. During highway driving tests, the Long Range model achieved around 94 MPGe, slightly lower than the official estimate, while the Performance model recorded 98 MPGe. These figures, while reflecting the energy efficiency of the Model Y, highlight the importance of considering individual driving patterns and conditions for a more accurate understanding of real-world fuel economy.

Chapter 2: Design and Build Quality

Exterior Styling

The exterior styling of the 2024 Tesla Model Y maintains Tesla's minimalist and futuristic design language, characterized by clean lines and a sleek silhouette. The SUV's high-riding body style, inherited from the Model 3 platform, adds an element of versatility to the aesthetic appeal. While the Model Y doesn't feature the distinctive Falcon Wing doors of the larger Model X, its design remains contemporary and aerodynamically optimized. The absence of a traditional front grille, a hallmark of electric vehicles, contributes to the vehicle's streamlined and distinctive appearance. With a focus on both form and function, the Model Y's exterior design not only enhances its overall efficiency but also solidifies its presence as a modern and forward-thinking electric SUV.

Interior Comfort and Space

Step inside the 2024 Tesla Model Y, and you'll find a spacious and uncluttered interior that prioritizes comfort and functionality. The cabin design is a testament to Tesla's commitment to simplicity, featuring a large touchscreen at the center of the dashboard, which consolidates various controls and functions. The optional all-glass roof further enhances the sense of openness, providing an airy ambiance for both front and rear passengers. The Model Y's interior accommodates five passengers with generous headroom and legroom in both the first and second rows. However, the optional third row, available for an additional cost, is suitable primarily for smaller passengers due to limited space. Despite some ergonomic compromises, the overall interior design emphasizes a contemporary and user-friendly experience, aligning with Tesla's commitment to innovative and modern vehicle interiors.

Cargo Capacity and Practicality

One of the standout features of the 2024 Tesla Model Y is its practicality, evident in its impressive cargo capacity and versatility. The high-riding SUV body style contributes to a more accommodating cargo area compared to the Model 3 sedan. The rear seats can be folded to create a flat load floor, providing ample space for transporting larger items. The absence of a traditional gasoline engine allows for additional storage space in the front trunk, often referred to as the "frunk." This dual-storage configuration enhances the Model Y's suitability for those with active lifestyles or frequent cargo-hauling needs. While the optional third row may limit the cargo space when in use, it adds flexibility for families or individuals requiring occasional seating for additional passengers. Overall, the Model Y strikes a commendable balance between passenger comfort and cargo-carrying capability, making it a practical choice for a variety of lifestyles.

Infotainment and Connectivity

At the heart of the 2024 Tesla Model Y's interior is the central touchscreen, serving as the command center for infotainment and connectivity features. The large display not only houses the entertainment system but also serves as the primary interface for various vehicle controls, including climate settings and navigation. Tesla's infotainment system offers a user-friendly experience, featuring crisp graphics and responsive touch controls. The Model Y inherits the entertainment functions from the Model 3, providing embedded apps for streaming services like Netflix, Hulu, and YouTube, as well as video games to keep occupants entertained during charging stops. While the reliance on touchscreen controls for nearly all secondary functions may pose a learning curve for some users, it contributes to the overall modern and tech-centric appeal of the Model Y's interior. Connectivity features include over-the-air software updates, ensuring that the vehicle stays up-to-date with the latest enhancements and improvements, showcasing Tesla's commitment to continuous innovation in both design and technology.

Chapter 3: Model Y Variants

Standard Range

The 2024 Tesla Model Y Standard Range variant represents the entry-level offering in the Model Y lineup, providing an accessible option for those entering the electric vehicle market. Expected to be priced around $46, 000, this variant offers a compelling balance of affordability and electric performance. Featuring rear-wheel drive, the Standard Range is designed for efficient commuting and daily use. While it may not boast the same acceleration figures as its Long Range and Performance counterparts, the Standard Range is expected to offer a respectable driving range, making it suitable for urban dwellers and those with moderate commuting needs. With its minimalist yet functional approach, the Standard Range appeals to buyers seeking an electric SUV without the additional features and performance enhancements found in higher trims, making it an economical choice for those making the switch to electric mobility.

Long Range

The Long Range variant of the 2024 Tesla Model Y caters to those seeking an extended driving range without compromising on performance. Priced around $51, 000, the Long Range introduces dual motors for all-wheel drive, enhancing both acceleration and overall driving dynamics. With an estimated driving range of 330 miles on a single charge, the Long Range model becomes an attractive option for those with longer commutes or planning frequent road trips. Its acceleration, while slightly behind the Performance model, still offers an impressive 0 to 60 mph time of 4. 4 seconds. The Long Range balances efficiency and performance, making it a versatile choice for individuals and families requiring a practical yet spirited electric SUV. With the dual-motor setup providing enhanced traction and stability, the Long Range variant adds a layer of confidence to the driving experience.

Performance Model

For enthusiasts seeking the pinnacle of performance in the 2024 Tesla Model Y lineup, the Performance model stands out as the top-tier offering. Priced at around $55, 000, the Performance variant takes electric SUV acceleration to new heights, achieving 0 to 60 mph in a blistering 3. 6 seconds. This model not only emphasizes speed but also introduces performance-oriented features, including 20-inch wheels, a lowered suspension, and a dedicated Track mode. The Performance Model caters to drivers who prioritize an exhilarating driving experience, whether it be on city streets or spirited drives on open roads. While the driving range is slightly reduced to 303 miles, the Performance model aims to deliver an unmatched combination of speed, agility, and dynamic handling, solidifying its status as the sports car of the Model Y lineup.

Choosing the Right Model for You

Choosing the right variant of the 2024 Tesla Model Y involves considering individual preferences, lifestyle, and driving needs. The Standard Range variant appeals to budget-conscious consumers, offering essential electric mobility without the bells and whistles. It's an ideal choice for those primarily using the vehicle for daily commuting in urban or suburban settings.

The Long Range variant steps up the game, providing an extended driving range and all-wheel-drive capabilities. This makes it suitable for a broader audience, including those with longer commutes or the occasional need for enhanced traction.

Enthusiasts and performance-oriented drivers may find the Performance model irresistible. With its sports car-like acceleration and performance enhancements, it caters to those who prioritize an exhilarating driving experience and are willing to sacrifice a bit of range for sportier features.

Ultimately, the right Model Y variant depends on the buyer's lifestyle, preferences, and budget. Each variant offers a unique balance of performance, range, and affordability, ensuring that there's a Model Y for every type of electric vehicle enthusiast.

Chapter 4: EV Technology

Electric Motor and Powertrain

At the core of the 2024 Tesla Model Y's impressive performance is its advanced electric motor and powertrain technology. The Model Y lineup features electric motors designed for efficiency and instant torque delivery. The Standard Range variant boasts a rear-wheel-drive configuration, while the Long Range and Performance models benefit from a dual-motor setup, providing all-wheel drive for enhanced traction and control. This dual-motor configuration not only improves acceleration but also contributes to the vehicle's stability, particularly in challenging driving conditions. Tesla's commitment to electric propulsion technology is evident in the seamless integration of these motors, delivering a driving experience that aligns with the brand's reputation for innovation and cutting-edge engineering. The powertrain, including the single-speed direct-drive transmission, ensures a smooth and responsive transition between speeds, enhancing both performance and efficiency.

Range, Charging, and Battery Life

The 2024 Tesla Model Y excels in addressing one of the primary concerns of electric vehicle owners—driving range. The Long Range model, in particular, boasts an impressive estimated range of 330 miles on a full charge, providing a significant buffer for various driving scenarios. The Performance model, while slightly reduced at 303 miles, still offers a competitive range. Tesla achieves these figures through advancements in battery technology and aerodynamics. The Model Y is equipped with a high-capacity lithium-ion battery pack, and the vehicle's aerodynamic design minimizes drag, contributing to overall energy efficiency. Charging the Model Y is made convenient through Tesla's extensive Supercharger network, allowing for rapid charging at dedicated stations. The vehicle supports both AC and DC charging, providing flexibility for different charging infrastructures. While charging times may vary, Tesla's commitment to improving charging infrastructure contributes to the overall feasibility and convenience of electric vehicle ownership.

The Model Y's battery life is supported by Tesla's comprehensive warranty coverage, ensuring confidence in the longevity and reliability of the electric powertrain.

Real-World MPGe and Fuel Efficiency

The Environmental Protection Agency's (EPA) MPGe (miles per gallon of gasoline-equivalent) ratings provide a standardized measure for comparing the efficiency of electric vehicles. The Long Range model of the 2024 Tesla Model Y is estimated to achieve 127 MPGe in the city and 117 MPGe on the highway, showcasing its efficiency in urban and highway driving scenarios. The Performance model, while slightly less efficient, still impresses with estimates of 115 MPGe in the city and 106 MPGe on the highway. However, it's essential to note that real-world driving conditions may yield variations from these standardized estimates. In highway driving tests, the Long Range model achieved around 94 MPGe, and the Performance model recorded 98 MPGe. These results highlight the impact of factors such as driving speed, terrain, and temperature on the Model Y's fuel efficiency. Nevertheless, the Model Y's overall efficiency underscores the advancements in electric vehicle technology, offering a compelling alternative to traditional internal combustion engines in terms of both performance and environmental sustainability.

Chapter 5: User Experience

Interior Features and Ergonomics

The 2024 Tesla Model Y prioritizes a sleek and minimalist interior design, focusing on features and ergonomics that contribute to a comfortable and intuitive user experience. The cabin is designed with simplicity in mind, featuring clean lines and a lack of physical buttons. The seating is both supportive and spacious, ensuring a pleasant ride for occupants. Tesla's commitment to an uncluttered interior extends to the controls, with most functions accessible through the central touchscreen. While this design choice may initially pose a learning curve for some users, it ultimately streamlines the cabin, emphasizing a modern and tech-centric environment. Interior features such as the optional all-glass roof contribute to an open and airy feel, enhancing the overall driving experience. Despite some ergonomic compromises, the Model Y's interior features and layout cater to users seeking a contemporary and user-friendly electric vehicle.

Touchscreen Controls and User Interface

Central to the user experience of the 2024 Tesla Model Y is the large touchscreen at the center of the dashboard, serving as the primary interface for various vehicle functions. This touchscreen controls everything from infotainment to climate settings and even the steering column angle. While this consolidated approach minimizes physical buttons and contributes to the overall sleek design, it also means that nearly all secondary controls are reliant on the touchscreen. Users may find adjusting settings, such as power mirror adjustments or steering column angle, to be less intuitive compared to traditional physical controls. However, the touchscreen is responsive and features a user-friendly interface, offering access to a range of functions and settings. Tesla's over-the-air software updates further enhance the touchscreen's capabilities, ensuring that users benefit from ongoing improvements and additional features, showcasing the brand's commitment to delivering a cutting-edge user interface.

Customization Options and Exterior Colors

The 2024 Tesla Model Y provides users with limited but meaningful customization options, allowing buyers to personalize their electric SUV to some extent. While Tesla's approach to trim levels is straightforward, with the Standard Range, Long Range, and Performance models, users can choose from a handful of exterior colors to add a touch of individuality. Customization options extend to choosing between different wheel designs, adding further aesthetic appeal. However, compared to some competitors in the automotive market, the Model Y's customization palette may seem limited. The available exterior colors often come at an additional cost, emphasizing Tesla's commitment to transparent pricing. Despite the somewhat constrained range of customization options, the Model Y's design remains modern and attractive, appealing to users who appreciate the sleek and timeless aesthetic of Tesla's electric vehicles.

Optional Third Row of Seats

For users with larger families or a need for occasional additional seating, the 2024 Tesla Model Y offers an optional third row of seats for an extra cost. While the third row is a welcome addition for expanding the SUV's passenger capacity, it's essential to note that the space is limited and best suited for smaller passengers. The two additional seats in the third row may not comfortably accommodate adult-size passengers, making them more suitable for children or short-distance travel. The ability to fold the third-row seats when not in use adds flexibility to the Model Y's interior, allowing users to prioritize either passenger space or cargo capacity based on their needs. The optional third row caters to users who value versatility and occasional seating options, making the Model Y a more family-friendly electric SUV.

Chapter 6: Safety Features

Tesla's Autopilot and Full Self-Driving Capability

Tesla's Autopilot and Full Self-Driving (FSD) Capability stand at the forefront of the 2024 Model Y's safety features, showcasing the brand's commitment to advancing autonomous driving technology. Autopilot, included as a standard feature, provides driver-assist capabilities such as adaptive cruise control, lane-keeping assist, and automatic emergency braking. While the term "Autopilot" might suggest full autonomy, it's crucial to understand that, despite its name, it does not render the vehicle fully self-driving. Tesla's Full Self-Driving Capability, available as an optional upgrade, introduces additional features such as Navigate on Autopilot, Summon, and Autopark. However, it's important to note that even with FSD, driver attention and intervention are still required, and the system does not make the vehicle fully autonomous. Tesla continually improves these features through over-the-air software updates, enhancing safety and functionality. The combination of Autopilot and FSD Capability contributes to the Model Y's reputation for cutting-edge safety technology, albeit with the understanding that these systems are designed to assist rather than replace the driver.

Driver-Assistance Features

The 2024 Tesla Model Y incorporates a range of driver-assistance features aimed at enhancing safety and reducing driver fatigue during long journeys. Standard across all Model Y variants, these features include automated emergency braking with pedestrian detection, lane-departure warning with lane-keeping assist, and adaptive cruise control with a lane-centering feature. These driver-assistance features work in tandem to create a comprehensive safety net, particularly on highways and well-marked roads. The lane-centering feature within the adaptive cruise control system helps maintain the vehicle's position within the lane, while automated emergency braking enhances collision avoidance capabilities. Tesla's commitment to providing these features as standard demonstrates the brand's dedication to prioritizing safety in their vehicles, setting a benchmark for the industry in integrating advanced driver-assistance technologies.

Crash Test Results and Safety Ratings

The safety of the 2024 Tesla Model Y is further underlined by its crash test results and safety ratings from prominent organizations. While official crash-test results from the National Highway Traffic Safety Administration (NHTSA) and the Insurance Institute for Highway Safety (IIHS) may vary, Tesla's vehicles, including the Model Y, have historically performed well in these evaluations. The NHTSA and IIHS assess aspects such as frontal crash, side crash, and rollover resistance, providing a comprehensive evaluation of a vehicle's safety performance. Tesla's vehicles have often received high marks in these assessments, contributing to the brand's reputation for producing safe and secure electric vehicles. Prospective buyers seeking detailed crash-test results and safety ratings can refer to the NHTSA and IIHS websites for the most up-to-date information, ensuring that safety remains a key consideration in their decision to embrace electric mobility with the Model Y. The combination of active safety features, driver-assistance technologies, and strong crash-test results positions the 2024 Tesla Model Y as a compelling choice for safety-conscious consumers in the electric SUV market.

Chapter 7: Pricing and Value

Price Range and Trim Options

The 2024 Tesla Model Y arrives with a competitive pricing structure, catering to a range of consumers seeking to embrace electric mobility. With an estimated starting price of around $46, 000 for the Standard Range, $51, 000 for the Long Range, and $55, 000 for the Performance model, Tesla positions the Model Y as a compelling choice in the electric SUV market. The diverse trim options cater to varying preferences and priorities. The Standard Range variant targets budget-conscious consumers looking to enter the electric vehicle space, offering essential features without compromising on electric performance. The Long Range model introduces all-wheel drive and an extended driving range, making it a versatile option for those with longer commutes or road-trip enthusiasts. The Performance model, with its emphasis on acceleration and sportier features, appeals to enthusiasts seeking a more dynamic driving experience. Tesla's transparent pricing, along with the clear distinction between trim levels, ensures that buyers can make informed decisions based on their specific needs and preferences.

Customization and Additional Costs

While the 2024 Tesla Model Y offers customization options, the range may seem limited compared to some traditional automakers. Buyers can choose from a selection of exterior colors and wheel designs to add a personal touch to their electric SUV. However, the majority of customization options come with an additional cost, aligning with Tesla's approach to transparent pricing. Optional features, such as the third row of seats, carry an extra charge, allowing users to tailor the Model Y to their specific requirements. It's essential for buyers to factor in these additional costs when configuring their ideal Model Y. Tesla's straightforward approach to pricing and customization aims to provide clarity to consumers, ensuring they understand the overall cost of their chosen configuration. While some competitors may offer a broader range of customization options, Tesla's focus on cost transparency contributes to a straightforward and hassle-free buying experience.

Comparisons with Other Electric SUVs

As the electric vehicle market continues to expand, the 2024 Tesla Model Y faces competition from other electric SUVs, each with its unique features and value proposition. Key rivals include the Ford Mustang Mach-E and the Volkswagen ID. 4. While the Model Y impresses with its driving range, performance, and Supercharger network, competitors offer alternative advantages. The Ford Mustang Mach-E, for instance, appeals to those seeking a balance between performance and a traditional SUV aesthetic. The Volkswagen ID. 4 emphasizes a spacious interior and a more affordable starting price. Choosing the right electric SUV involves considering individual priorities, whether it be driving range, performance, or overall value. The Model Y's pricing, while competitive, is worth evaluating in the context of its competitors to ensure that buyers find the best fit for their specific needs. Tesla's Supercharger network and Autopilot features contribute to the overall value proposition, but consumers should weigh these against the offerings of other electric SUVs in the market to make an informed decision.

Chapter 8: Ownership Experience

Warranty Coverage

The 2024 Tesla Model Y comes with a warranty package that reflects Tesla's confidence in the quality and durability of their electric vehicles. The warranty coverage aligns with Tesla's other models, including the Model 3, Model S, and Model X. The bumper-to-bumper warranty spans four years or 50, 000 miles, providing comprehensive coverage for the vehicle during its initial years on the road. Additionally, the powertrain warranty extends for eight years or 100, 000 miles, showcasing Tesla's commitment to the longevity of the electric motors and battery. This warranty package not only instills confidence in buyers regarding the reliability of the Model Y but also demonstrates Tesla's dedication to standing behind their products. While the warranty coverage is in line with industry standards, Tesla's ongoing commitment to over-the-air updates ensures that the Model Y remains up-to-date with the latest software enhancements, contributing to the overall ownership experience.

Maintenance Costs and Considerations

Owning an electric vehicle often translates to lower maintenance costs compared to traditional internal combustion engine vehicles, and the 2024 Tesla Model Y is no exception. With fewer moving parts and no need for oil changes, electric vehicles typically incur lower maintenance expenses over their lifespan. While routine maintenance tasks such as tire rotations and brake inspections are still necessary, the absence of complex engine components simplifies the maintenance process. Tesla, however, does not offer complimentary scheduled maintenance, meaning owners are responsible for these routine service costs. The long-term savings associated with electric vehicles, including reduced fuel costs and lower maintenance requirements, contribute to the overall value proposition of owning a Model Y. While the initial purchase price might be higher than some gas-powered SUVs, the potential for lower ongoing maintenance costs enhances the appeal of the Model Y as a cost-effective and sustainable transportation solution.

Long-Term Reliability

Assessing the long-term reliability of the 2024 Tesla Model Y involves considering various factors, including the vehicle's track record, technological advancements, and feedback from existing owners. While Tesla's electric vehicles have garnered praise for their innovative features and performance, there have been occasional reports of reliability concerns, including issues with build quality and electronic components. However, it's important to note that Tesla continually addresses such issues through over-the-air updates and improvements in manufacturing processes. The 2024 Model Y benefits from Tesla's cumulative experience in producing electric vehicles, potentially leading to enhanced reliability over time. Real-world experiences from Model Y owners, available through online forums and reviews, can provide valuable insights into the day-to-day reliability of the vehicle. As electric vehicles become more prevalent, Tesla's commitment to continuous improvement and the Model Y's incorporation of the latest technologies contribute to a positive outlook on its long-term reliability. Prospective buyers may find reassurance in Tesla's proactive approach to addressing issues and the positive experiences of a growing community of Model Y owners.

Conclusion

Final Verdict

In conclusion, the 2024 Tesla Model Y asserts itself as a compelling option in the electric SUV market, combining Tesla's renowned electric technology with the practicality of an SUV. The Model Y's standout features include an impressive driving range, swift acceleration, and the convenience of Tesla's Supercharger network. While its driving dynamics may not match the sportiness of the Model 3, the Model Y prioritizes spaciousness and versatility, making it a suitable choice for families or those with an active lifestyle.

The choice of trim levels—Standard Range, Long Range, and Performance—allows buyers to tailor their Model Y to specific needs, whether it's a balance of performance and range or an emphasis on acceleration and sporty features.

However, the reliance on touchscreen controls and occasional concerns about build quality might be factors for some potential buyers to consider.

Pros and Cons

Pros:

- **Impressive Driving Range:** The Model Y excels in offering a competitive driving range, particularly in the Long Range variant, making it suitable for various driving scenarios.

- **Swift Acceleration:** Both the Long Range and Performance models showcase Tesla's commitment to high-performance electric vehicles, delivering impressive acceleration figures.

- **Spacious Interior:** The Model Y prioritizes interior space, providing ample room for both passengers and cargo, making it a practical choice for families.

- **Autopilot and Full Self-Driving Capability:** Tesla's advanced driver-assistance features contribute to safety and offer convenience during long highway drives.

- **Supercharger Network:** The extensive Supercharger network enhances the Model Y's practicality by providing rapid charging options for long-distance travel.

Cons:

- **Not as Fun as Model 3:** The transition to an SUV body style results in the Model Y being less engaging to drive compared to its sedan counterpart, the Model 3.

- **Subpar Build Quality:** Some users have reported concerns about build quality, with occasional issues in fit and finish, which is less common among competitors.

- **Overly Reliant on Touchscreen Controls:** The reliance on the central touchscreen for most controls may be inconvenient for users accustomed to traditional physical buttons.

Closing Thoughts on the 2024 Tesla Model Y

The 2024 Tesla Model Y maintains its status as a compelling electric SUV option, building on the success of its predecessors. With a focus on driving range, performance, and the practicality of an SUV, the Model Y appeals to a broad audience of electric vehicle enthusiasts. The ongoing advancements in Tesla's Autopilot features and the potential for over-the-air updates contribute to a sense of future-proofing, ensuring that the Model Y remains at the forefront of electric vehicle technology.

While the Model Y is not without its challenges, including concerns about build quality and the transition to touchscreen controls, its strengths in driving range, acceleration, and overall versatility position it as a strong contender in the ever-evolving landscape of electric SUVs. As electric mobility continues to gain traction, the 2024 Tesla Model Y stands as a testament to Tesla's commitment to innovation, providing a glimpse into the future of sustainable and high-performance transportation. Prospective buyers looking for a well-rounded electric SUV with advanced technology and an extensive charging infrastructure will find the Model Y to be a worthy and forward-thinking choice.